AVIATION PHOTO ALBUM

LEE & AILI HIGBEE
1468 N. LARKWOOD SQ.
FT. MYERS, FLA. 33907

PSL Patrick Stephens, Cambridge

compiled and edited by

MICHAEL JF BOWYER

AVIATION PHOTO ALBUM

First published—1978

British Library Cataloguing in Publication Data

Aviation photo album.
1. Aeronautics—History—Pictorial works
I. Bowyer, Michael John Frederick
629.133'09'04 TL515

ISBN 0 85059 297 6

Design by Tim McPhee

Text set in 9 on 10pt Univers
by Stevenage Printing Limited.
Printed in Great Britain on 100 gsm
Buccaneer Matt coated cartridge and bound by
The Garden City Press of Letchworth
for the publishers Patrick Stephens Limited,
Bar Hill, Cambridge, CB3 8EL, England.

CONTENTS

INTRODUCTION 6
WORLD WAR I 9
BETWEEN THE WARS 15
WORLD WAR 2 33
THE POST WAR PERIOD 68

INTRODUCTION

In December 1965 a new feature, 'Photopage', was introduced into *Airfix Magazine*. The aim was to present, regularly, a selection of photographs of aircraft and, in particular, previously unpublished illustrations. Readers were asked to submit such photographs and the contributions came in abundance. Many of these photographs were exceedingly interesting, usually rare, and indeed 'Photopage' proved popular, and was for some time a very regular feature. Its following was truly world-wide.

Amongst the welter of material came photographs of such unusual items as the Blackburn Skua in service, and one published illustration depicted a Short Stirling in Russian markings.

Retention of any one feature in a magazine for a very long period is not entirely satisfactory, however popular it is, for a fresh approach maintains a lively product. Wide readership has certainly been a satisfying feature of *Airfix Magazine*. This has been reflected in the contributions to 'Photopage' for the contributors ranged from as far away as Argentina, northern Canada, South Africa, Portugal, Korea, New Zealand, Australia and even Eastern Europe. It proved beyond any doubt that interest in refreshingly new and unusual photographs literally knew no bounds.

As years passed the volume of material submitted became quite embarrassing, and was far more than the magazine could ever carry. Space being at a premium thoughts were directed upon how to make these excellent contributions available to the many who enjoy them. It was therefore decided to bring some of them together in a book. Our hope is that the popularity of this idea might encourage more fine contributions which, if the book proves popular, can lead to others since clearly the type of photograph contained herein must have wide appeal.

Submitted material has spanned the entire history of aviation from balloons to the latest jet aircraft, and as can be seen from this volume new material from World War 1 has appeared.

It will be seen that the book's contents fall into four periods roughly reflecting the obvious interests of the readership of *Airfix Magazine*, and whose leanings are clearly towards the period 1939-1945. It opens with a selection of photographs of the period 1914-1918, the quality of which varies. Sad to say, often the most interesting pictures are of poorer quality than one would wish, but this has not been the criterion for inclusion or otherwise. Selection was on the basis of freshness and rarity.

Photographs of the inter-war period have appeared in some numbers and, whilst this is a period of less appeal to many readers, the specialist will doubtless find it of interest. The quality of some of the pre-war pictures has been exceptional, for this was a time when glass plates were in widespread use although the choice of orthochromatic film renders some of the tones unusual suggesting strange colouring.

In the popular 1939-1945 selection the aim has been to include the less usual, again at the expense sometimes of quality. One of the strange aspects of this section is the absence of many of the best known wartime aircraft. One surely never ceases to be surprised at the relative absence of photographs of Spitfires in service, of Blenheims, Wellingtons, Halifaxes, Hampdens and many more. Photographs abound of pilots standing by their aircraft, but of complete machines there is sad paucity. Do they exist? Possibly, probably, but they remain rare. As the years pass the likelihood of many being lost for all time increases, so it is surely up to those of us interested in such historic material to make every effort to see that it is saved. That the material does exist is suggested by such photographs as Len Bachelor's submission of a long-range Spitfire of 66 Squadron.

In the post-war selection concentration had tended to be towards the immediate post-war years when RAF Transport Command was strong and wartime aircraft were still to be seen in plenty. It is now over 30 years since the war ended, and many unusual aircraft have graced the skies. Thus, the selection has been brought up to date to show the great contrast between the Sopwiths and the Lightning, although reflecting our obvious readers' interest such items are few.

Captioning has been kept to bare essentials for it is clear that it is the photographs which must claim the attention. Thus, type, name and place with likely dating has been concentrated upon.

Inevitably, the photographs are in black and white. The introduction of colour would mean a great increase in costing, and it is rarely that aeroplanes appear in print in their true colours. A highly lit aeroplane may make a pretty picture, but the great variations in colour temperature in all colour material often presents a strange illustration. In any case colour photography centres upon recent years and the subjects often lack the fascination of the older types of aeroplanes we all enjoy seeing in print.

Let it not be thought, however, that colour photography is merely an idea of the 1950s. One reader produced a colour slide of the Short 'Knuckleduster' flying-boat taken at Felixstowe in the 1930s—and the colour was good even thought it showed the trade mark of Dufaycolour. This merely goes to show what is available for the ardent seeker.

What has never ceased to amaze us has been the patience shown by readers who have waited long for photographs to be published. It has always been our intention to publish as many of the contributions as we can, and this book is one way of speeding up the process.

Equally amazing has been the number of correspondents who have sent along photographs and not included a name and address—or even a covering note. If you are one of these you may find that we have included a few such contributions, and hope that you will contact us.

The fascination of collecting photographs of aircraft veritably exploded in the 1960s, before which photographers were far fewer. The quality of their work was nevertheless high, at a time when suitable cameras were less abundant. Photographs of high resolution could not be achieved by many and definition was often poor. Full marks, then, to those who have found such splendid material from years past.

Without doubt the advance of 35 mm cameras has altered the entire spectrum of aircraft photography. It has brought photography of high standard within the grasp of many as any visitor to any present day display can testify. Prior to 1953 photography of RAF aircraft was rarely permitted. After the Odiham review it was banned again and it was some months before it was finally lifted at public events. Much historic material was inevitably lost. The American attitude was less restrictive, fortunately, as can be seen from some of the items herein.

It is to be hoped that, if you are an aviation photographer, you will enjoy looking at the efforts of others, some from over a half century ago, and that it will encourage you to seek for the past in many a likely place. Bearing in mind the security aspects of photographing military aircraft it is hoped that from now on every effort will be made to retain worthwhile photographic records for posterity.

Do not neglect the civil scene either, for there is so much therein that is highly photogenic. Who could not fall for a picture now of an Empire flying boat or a Hannibal?

Good shooting!

Michael J.F. Bowyer

Michael J.F. Bowyer
Cambridge, October 1977

A1590

WORLD WAR I

Previous page No details are available concerning this Martinsyde Elephant A1590, but the picture may have been taken in North Africa in 1917 (*W.J. Beeby*).

Left Sopwith Camel B6208, squadron number uncertain, but possibly No 64 Squadron (*H. Constable*).

Below Bristol F2B B1112 (Falcon III engine) of No 6 Squadron (*H. Constable*).

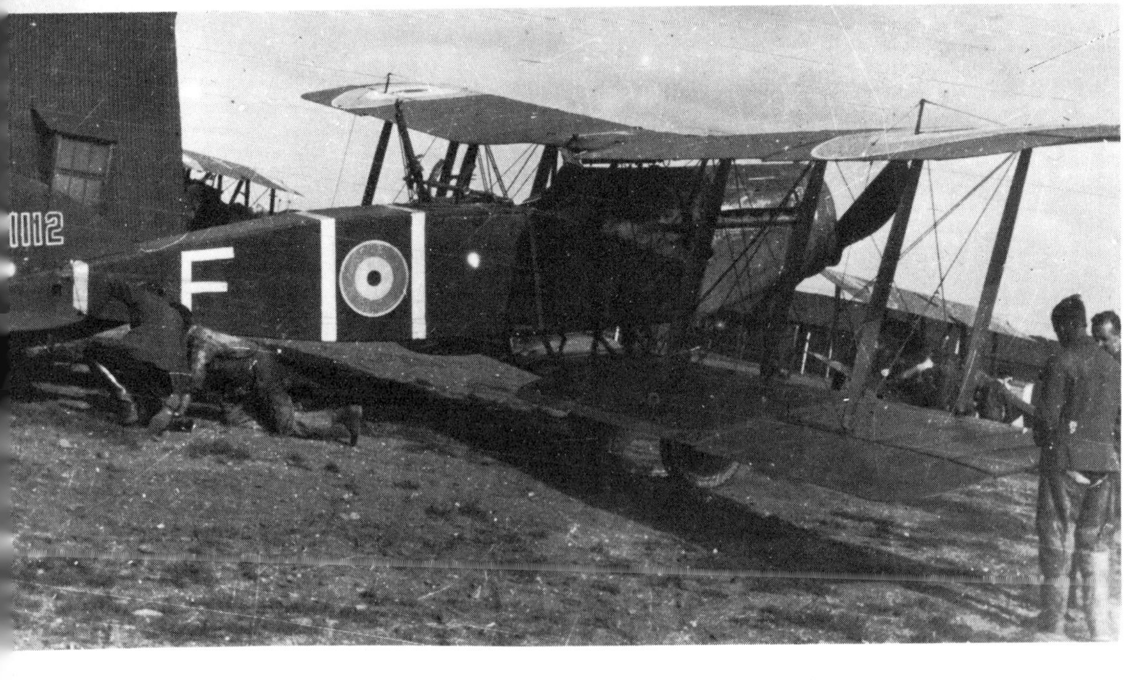

Right Air-to-airs taken in World War 1 are uncommon, but distantly here is a Morane monoplane (*H. Constable*).

Below Sopwith 1½ Strutter '?652', probably of No 64 Squadron (*H. Constable*).

Right An Armstrong-Whitworth FK 8 F7411, date and place not known. The contributor failed to give us a name or address!

Left Where and when this Albatross DV was captured is unknown (*M. Kent*).

Left Other than the serial N2369 nothing more is known about this Fairey Campania (*R. Daniels*).

Left B6625 is said to have been an RT 1, but is more likely an RE 8. Photographed at London Colney (*R.S. Humphreys*).

Right From an American reader came this fine shot of an LFG Roland C-11 'Walfish' fighter of 1915-1917 vintage (*J.E. Lopo*).

F
7411
F.7411
19

500

BETWEEN THE WARS

Above Blackburn Baffin K3548 suspended by a crane of HMS *Courageous* (*N.B. Leicester*).

Previous page The certain end of Vickers Virginia K2670 'K' of 500 Squadron, one of the last 'Ginnys'. September 16 1933, at Brooklands (*R.N. Eason*).

Above right Oh for a walk round that deck now with a Kodachrome! Baffins roaring away on HMS *Courageous* (*N.B. Leicester*).

Left A Fairey Flycatcher of 401 Flight taking off from HMS *Furious* in 1932 (*R.J. Maclachlan*).

Left Fairey IIID S1010 over Hong Kong around 1927-28. Appears to carry '47' on the fuselage band (*V.W. Clarkson*).

Right Into action—a Blackburn Ripon II practising torpedo attack. The swastika on the wheel is simply a good luck symbol (*N.B. Leicester*).

75
62

S
1265
75
S.1265

Hold on chaps, we don't have many of them! Baffin '72' on the bow of *Courageous* (*N.B. Leicester*).

You should be more careful, otherwise Baffins end up like '73' (*N.B. Leicester*).

Fairey IIIF Mk IV in the hands of No 24 Squadron (*contributor not named*).

A very fine shot of Fairey Seal K3481, coded '726' (*A.W. Feist*).

The prototype Fairey Hendon K1695 at Boscombe Down in 1932 (*P.A. Ruff*).

An early photograph of the Vickers B 19/27 bomber (*S.C.M. Davis*).

Bristol Bulldog IIA K216?, almost certainly of No 19 Squadron, from an anonymous contributor.

All the majesty of a 'boat'— Singapore III K4580:1 of No 230 Squadron (*T. Jones*).

Where haunt the Harriers now, Hawker Hart (Special) K3142 of 3 FTS awaits a pilot at Wittering (*J.H.M. Richardson*).

All the flavour of the 1920s! RAF ground crew by the nose of a Vickers Virginia of No 9 Squadron (*R. Gascoigne*).

K
3848
K3848
K3845

Left Lined up at Manston, three Hawker Harts of No 500 Squadron. They are K3848, K3845 and K3816 (*A. W. Feist*).

Below left Photographs of Avro Prefects are rare. K5068 shown here at Manston was serving with the Air Navigation School there (*G. Stanley*).

Right H2986, an Avro 504, in the hands of the RNZAF (*N. B. Leicester*).

Below Hawker Osprey III float-plane K3636, at Cowes? In the distance a Saro Cloud (*A. W. Feist*).

Left Hawker Hart (T) K6426 of the Straits Settlements Volunteer Air Force tags along with some Vildebeests to greet the Governor (*contributor not named*).

Right Vickers Vildebeeste NZ105 of the RNZAF (*J. Lumley*).

Below Similar, but a general purpose variant, was the Vickers Vincent, NZ341 seen here. Note the camouflaged upper surfaces (*J. Lumley*).

Below right Beautiful, beautiful! A fine shot of a Fairey Gordon in service with the RNZAF as a target tower (*J. Lumley*).

NZ
105
NZ105

NZ
631
NZ 631

K
6335
K6335

B

Far left K6335, a Vincent of No 55 Squadron (*D.A. Needs*).

Left Far removed from pressurised refuelling! A Vincent of No 55 Squadron during a refuelling operation (*D.A. Needs*).

Right Over an Arab settlement, K6335 flies before transfer to the RNZAF (*D.A. Needs*).

Left No photo collection can be complete without a Valentia. K3165 here of No 70 Squadron instances the old faithful crossing Baghdad (*D.A. Needs*).

Right A Klemm trainer in Finland in 1936—no further details known (*Esa Savola*).

Right Heinkel He 112 D-IEFC—no further details known about the photograph except that it is said to date from 1936 and was taken in Finland (*Esa Savola*).

Right Among the interesting photographs received have been some of pre-war Czech aircraft, like this Letov S-231 fighter (*Jaroslav Zazvonil*).

Left Finnish Air Force Viima trainer photographed in 1936 (*Esa Savola*).

Left One of the loveliest pre-war sights, a Short C-Boat G-ADUW 'Castor' (*L.W. Cooper*).

Left 'Castor' beached. Taking a boat ashore was a major operation (*L.W. Cooper*).

Right Rare indeed, a shot of two Finnish Air Force Blackburn Ripons in 1936 (RI 129 and RI-155) (*Esa Savola*).

-129
RI-155

Left The Czech fighter shown here is the Avia BH-33L (*P.M. Bradley*).

Right Perhaps more than any other type the Bristol Blenheim represents the changeover to monoplanes in the RAF. L4924 is seen here being refuelled at Sharjah in June 1939 (*W.A. Gatehouse*).

Right Blenheim 1 L4820 photographed before delivery to No 55 Squadron (*Alex Vella*).

Below left A fine shot of Vickers Wellesley K7783 of 45 Squadron at Habbinya in 1937 (*D.A. Needs*).

Below No doubt where this hails from. Ju 52 3/m D-AMYY 'WILHELM SIEGERT' on a pre-war visit to Newtownards, Co Antrim (*J.N. McFarland*).

WORLD WAR 2

An early Heinkel He 111, possibly an He 111F, which crash-landed in Czechoslovakia (*Jaroslav Zazvonil*).

Previous page A Heinkel He 111H N + CO said to be attacking shipping in the Atlantic on November 4 1942 (*W. Skrzyporak*).

The how, where and when on this Bucker Bü-131 is not known but it was probably photographed in Czechoslovakia (*W. Skrzyporak*).

The prototype Fairey Battle had a streamlined spinner on its propeller, removed due to vibration problems. Yet within this line-up of 142 Squadron Battles in France in early 1940 at least two have spinners (*A.F. Woollett*).

Lisbon Airport before the storm broke—A German DC-3 D-AAIF is in the foreground, with DC-2 EC-AAB beyond and a Portuguese Lockheed to the right (*C.B. Fonseca*).

Above Captured Amiot reconnaissance aircraft at Abbeville in 1940 (*A.R. Chapman*).

Above left A clear shot of a Ju 86E SE + PX (*Jaroslav Zazvonil*).

Right Heinkel He 111H 5J + CH which was shot down on May 14 1940 at Yselstein, Holland (*B.W.C. van Albeslo*).

Left A most interesting shot of a German floatplane base showing Heinkel He 59s. No details known (*D. Morrison*).

Right Heinkel He 59 taxiing, but again no details are known (*Jaroslav Holeček*).

Above One of the batch of Curtiss Hawk H-75A-6 fighters for Norway in 1940. The Hawk 75 had a most diverse career, and one well worth study for so little has been fully authenticated (*Bjorn Olsen*).

Left A Ju 52 3/m SE+DM said to have been photographed in Russia in 1943 (*A.R. Chapman*).

Left A Heinkel He 118 photographed in France in 1940 with camouflage netting obscuring the markings (*A.R. Chapman*).

Above right Junkers Ju 52 3/m SE + 1M of Flugzeugführerschule (B)8 which forcelanded in Sweden with engine trouble in April 1940 (*N. Stewart*).

Right A Junkers W-34 BB + HC on its nose in Czechoslovakia (*W. Skrzyporak*).

Left A close-up of a Heinkel He 114 floatplane (*E. Savola*).

Below A Heinkel He 111H being prepared for action over Britain during the 1940 daylight campaign. Someone has painted out the Gruppen insignia (*W. Skrzyporak*).

Right What a pity the code letters on this Spitfire have been painted out! (*D. Morrison*).

Below In the Russian snow, a Heinkel He 111H taxies in (*W. Skrzyporak*).

Left Heinkel He 111 undergoing maintenance in a French hangar in 1940 (*A.R. Chapman*).

Right Hats off, chaps, it's one of ours! A Bf 109F returns to its airfield on the Eastern Front on July 29 1942 (*W. Skrzyporak*).

Left Heinkel He 118, possibly CE + BO, with a captured Russian SB-2 in the distance, in France (*A.R. Chapman*).

Below Heinkel He 114 VR + BH and a Dornier Do 18 during the Russian campaign (*E. Savola*).

Weapons being prepared for use by Ju 88s for an attack on Malta on August 14 1942 (*W. Skrzyporak*).

A shot of a Ju 88A, possibly in the Balkans (*W. Skrzyporak*).

More Junkers Ju 88s believed to have been photographed in the Balkans (*W. Skrzyporak*).

Junkers Ju 52 3/m H4 + AA photographed after attack in North Africa. The fin number appears to be 8655 (*J.A. Vella*).

A line-up of Dornier Do 217Es. The fuselage carries some lettering on the Balkenkreuz (*W. Skrzyporak*).

The emblem of the 'Rammjäger' carried on a Messerschmitt Bf 109 (*Vanhaezebrouck Robrecht*).

Reputed to be Me 210s of the Hungarian Air Force over Budapest in 1944 (*contributor not named*).

Wavy line camouflage can be seen on the nose of this Ju 88 of KG 54 (*Vanhaezebrouck Robrecht*).

Left A Ju 88, possibly Werk Nr 0886652 (866 on nose) after capture at Bardufoss, Norway, in May 1945 (*W.J. Gatehouse*).

Right The first attempt to extend the Spitfire's range was made by fitting a long-range tank on the port wing leading edge. LZ:B of 66 Squadron flying from Perranporth in July 1941 has this feature (*L.J. Bachelor*).

Left Mixed types of Ju 88s lined up at Bardufoss in May 1945, 866, subject of the previous photo, being nearest (*W.J. Gatehouse*).

Below A late war photograph of a FW 190 at its dispersal (*P.M. Bradley*).

Right The Allies found quite a large number of Me 262A jet fighters after the end of hostilies, among them this one. No details of capture known (*J.R. Kurrle*).

Below A captured Bf 110 late of NJG 1 photographed at Knokke in 1945 (*L. Hunt*).

Above Among the captured aircraft in 1945 were a number of Junkers Ju 188s which had been employed as anti-shipping aircraft (*J. Allyson*).

Above left 'Jolly Roger', a Lancaster of 463 Squadron with 59 operations recorded (*M. Kiddy-Perry*).

Left Nose art work on Lancaster V—'Victorious' of 463 Squadron. It has 31 raids and one enemy fighter to its credit (*M. Kiddy-Perry*).

Left Wellington KX:E-L7788 was forced down in Holland during a Berlin raid of September 23/24 1940. Photo graphed at Voorburg (*A. Stolk*).

Above Lancaster LL847 coded PO:D (Dull Red codes outlined yellow) served with 463 Squadron and was shot down during a Munich raid of December 17/18 1944 (*A. Tough*).

Right Other than the letters PO:U no individual identity is evident for this Lancaster of 463 Squadron. On the nose can be counted 104 sortie totals and the remark 'Goodbye old faithful' (*A. Tough*).

Right Beaufighters of a South African squadron flying in the Middle East, K5 nearest, B5 beyond (*Alan Green*).

Above left A shot of an Airspeed Oxford L4623 of the Royal Air College, Cranwell, with yellow extended up the fuselage side and appearing black on ortho film (*P. Connolly*).

Left Beaufighter 10 RD8?? at Habbaniya carrying a ferrying tank. Photographed July 1946 (*P. Glazier*).

Below Hands up anyone who would like to walk around this gathering . . . with a colour film handy! Lancasters during post-war trooping and POW retrieval, at Pomigliano near Naples (*M. Drackett*).

Left A Mosquito NF XIII night intruder (MM617?) of No 29 Squadron (*A.J. O'Connor*).

Right Mosquito FB VI HR526 marked with the SNAKE lettering intended to ensure fast transit to the Far East. In the hands of 7045 Servicing Flight, and serving with 45 Squadron (*C.P. Gunner*).

Right Tiger Moth T8023 '3' possibly photographed in Southern Rhodesia (*J. Brindley*).

Below Most wartime fighter squadrons held a Tiger Moth or a Magister for training and communications duties. L8054 is shown here at Andreas, Isle of Man, on January 10 1942 when on the strength of No 457 Squadron (*Frank F. Smith*).

Above Two Oxfords, AS690: 79 and BM720:76, flying over Southern Rhodesia. Overall yellow, black trim (*N.R. Hockin*).

Left Anson DG874 '30' in overall yellow approaching its training base in Canada in 1943 (*S. Mackay*).

Left RCAF Fleet Fort of No 3 Wireless School, Winnipeg, photographed on November 20 1943. Yellow overall (*R.B. Holswich*).

Above A most unusual shot of an Avro Anson. DJ549 served at the Air Observer's School RAAF, Mount Gambier, South Australia, in September 1944. Finish was overall green on the upper surfaces, light blue on the under surfaces, letters and numbers white (*M. Ryan*).

Right A Fairchild Cornell of the type extensively used in Canada for training. FV576 is shown here (*R. Staton*).

Right Overall yellow adorns Oxford HN607 '31' at Kumalo, Southern Rhodesia in 1944. Letters and numbers are black (*K. Jackson*).

Left Two Liberator VIs of No 104 Squadron in the Middle East. The nearest has the letters 'BL' painted out from its former days with 40 Squadron (*M. Barnes*).

Left An early photograph of a Boston III in the Middle East in 1942, pictured at Bilbeis, Suez Zone (*R. Kenway*).

Below Cessna Crane 7671 of the RCAF, in use for the Empire Air Training Scheme in Canada in 1943 (*S. Mackay*).

Right Westland Lysander L6859 served with No 225 Squadron in Britain January-August 1940. It was shipped to India in June 1941 and struck off charge there in July 1944 (*J. Brindley*).

Centre right One of the specially modified Spitfire Vs used for photo-recce work in the Middle East (*J. Brindley*).

Bottom right In Britain Fairchild Argus Is and IIs were extensively used during the war by the Air Transport Auxiliary. In the Middle East the Ranger-powered Argus III saw widespread service. A:Z-KK439 is shown here taxiing at Treviso in April 1947 (*S. Race*).

Below Art work 'Blue Devil' on the nose of a Hurricane IIC of No 11 Squadron in the Far East (*T. Large*).

Left A Bell P-39D Airacobra at Tripoli in 1942. Fin serial is 24742. No 24510 stands to the rear (*R. Staton*).

Right Lockheed Ventura Vs were used for weather reconnaissance in Britain and coastal patrol in the Middle East where later they were used as transports. All-silver FP598 shown here was photographed at Palam, India, in 1947 (*S.J. Lewis*).

Left Hurricane IID HV667 in Libya during March 1943 (*R. Staton*).

Below right An interesting shot of a Wellington Mk XIII of No 294 Squadron at Aqir, Palestine, in April 1945. Possibly ME941 or JA574 (*R. Kirby*).

Below Liberator GR VI of No 224 Squadron, possibly KH146, photographed in 1945 (*J.R. Kurrle*).

FP598

Left Some RAF Dakotas, particularly overseas, had disruptive camouflage on upper surfaces like FL507 here seen in India (*A. Mynett*).

Left Blackburn Skua L2833 at Andreas, Isle of Man, in 1941 (*Frank F. Smith*).

Below Although photographed in 1947, Spitfire XIV TZ114 still bears wartime markings. Photographed at Palam (*S.J. Lewis*).

Above Catalina III amphibian FP527 at Woodhaven(?). Joined 330 (Norwegian) Squadron June 17 1942 and operated from Woodhaven, and from Reykjavik from August 1942. In November 1942 passed to 131 OTU where it stayed until August 1944 (*Knut Arveng*).

Right A Martlet of 888 Squadron FAA, taking off in December 1942 wearing US markings to assist identity during operations off north-west Africa (*J.E. Young*).

Two Harvards with FE877 '16' during wartime flying training. No details of unit involved (*J.L. McDonald*).

A line-up of Fairey Firefly 1s in New Zealand, MB504:S nearest (*M.D.H. Hamilton*).

The *real* Messenger RG333 with the real General Montgomery photographed in 1945 (*J.R. Kurrle*).

Following page Venturas of the RNZAF during operations over the Pacific (*J. Lumley*).

NZ2522

"BEAU-GUNSVILLE"
DU-1
A8-186

Left A very clear photograph of a New Zealand Grumman Avenger, NZ2522 (*J. Lumley*).

Right Beaufighter A8-95 of No 5 OTU, RAAF, in March 1945. Finish dark green/light grey. Note the extended intakes above cowlings (*M. Ryan*).

Right An early wartime shot of a Curtiss Tomahawk in north Africa, serial unfortunately impossible to read (*M.J. Allen*).

Below left Beaufighter DU:I-A8-186 'Beau-Gunsville' undergoing servicing in Australia (*P.N. Anderson*).

Below A Lockheed Lodestar in use as a RNZAF transport (*J. Lumley*).

THE POST WAR PERIOD

ROYAL NAVY
A
DANGER

Previous page A splendid shot of a Westland Wessex HU 5 XS492 of No 848 Squadron, flying off Steamer Point, Aden (*N. McTeer*).

Left In the post-war period the Avro York soon became the mainstay of RAF Transport Command whose badge appears on the nose of MW294. Photographed at Sharjah in 1947 (*S.J. Lewis*).

Right The Duke of Gloucester's York 'ENDEAVOUR' at Bougainville in 1945 (*L. Cutler*).

Below right Old Tigers never die, they don't even fade away. These two were used by the Norwegian Air Force (*Bjorn Olsen*).

Below A rare sight, most surely, a RNZAF Miles Aerovan NZ1751 (*J. Lumley*).

Above For fast courier flying the RAF also used a few Avro Lancastrians like VM725 seen here at Changi in 1946 (*P. Finch*).

Left Heavy twin conversion flying was undertaken for a time using Bristol Buckmaster trainers, one of which is seen in this photograph at Habbaniya in November 1946 (*P. Glazier*).

Above Among the specialised Avro Lincolns used by special Schools after the war was RF532 'Thor II' of the EAAS, seen here at Habbaniya in April 1947 (*P. Glazier*).

Right Shorter range flying was by the trusty Douglas Dakota IV, like NU:V of 242 OCU seen here (*R. Barratt*).

Left Soon after the war a high speed courier service to Europe was flown by Mosquito 25s of 162 Squadron, Blackbushe. KA937 is seen at Kastrup (*W. S. Grigg*).

Below Vultee Vengeances were used after the war as target-tugs, unless like this machine they were soon scrapped. Photo taken at Kemajoran, Java, 1946 (*P. Finch*).

Right Mosquito PR34 PF677 of No 13 Squadron, in silver finish, at Seletar in 1952 (*J.E. Badcock*).

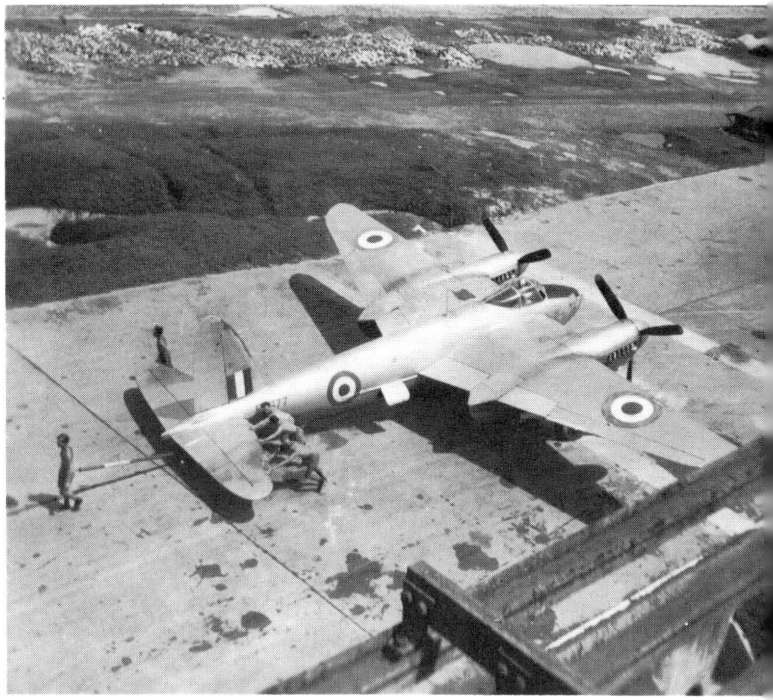

Below Wellington Xs were converted post-war into trainers and a few in the Middle East into transports. NA745 was one of these, seen here in Egypt (*P. Simms*).

Above In the post-war RNZAF P-51 Mustangs soldiered on, like NZ2406 (*contributor not named*).

Above right The French Navy acquired Lancasters like '15' here, photographed at Changi in November 1962 (*P.G. Ford*).

Left The usefulness of the Harvard extended long after the war and AJ628 of No 22 SFTS Thornhill, Southern Rhodesia, seen flying in July 1945, soldiered on (*G.M. O'Connell*).

Left A number of Ansons were exported but soon fell from favour like this machine of the Iraqi Air Force seen at Habbaniya in 1950 (*D.R. Nicholas*).

Right Another such Harvard was EX379 'J' of the Cranbourne Training School in Rhodesia (*D. Peace*).

EX379

The light bomber force in BAFO relied upon the Mosquito B 35, like EV:V-PF453 photographed in Italy in 1946 (*M.J. Smith*).

Many a Station Flight held an Airspeed Oxford. NJ282 served at Buckeburg in 1949 (*J. Price*).

Backbone of the ground attack force soon after the war was the Hawker Tempest II of which PR656 is shown at Bhopal in May 1946 (*M. Harries*).

Twin-engine training was afforded by Mosquito T IIIs at 204 AFS like FMO:L seen here at Driffield in 1948 (*R.A. Hill*).

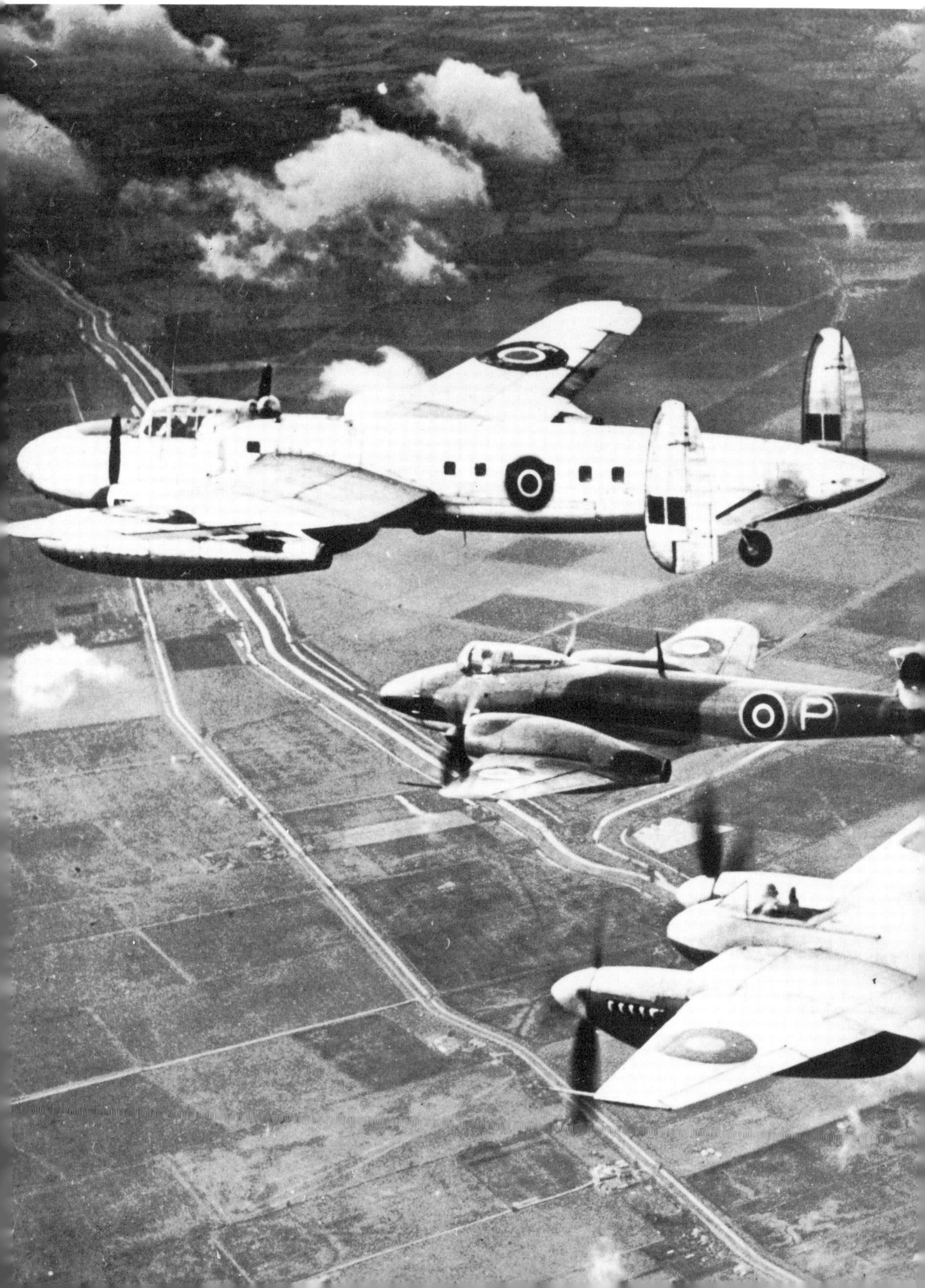

PX288

The magnificent lines of the Mosquito are well shown here, with FMO:L-VT606 nearest and camouflaged FMO:A-TV967 alongside (*R.A. Hill*).

Previous page The Rolls-Royce post-war test fleet. Hornet PX288 is nearest, then the Trent Meteor following the Nene Lancastrian with a Meteor 4 and Vampire beyond (*J.G. Burgess*).

A Mosquito PR 34 RG116 '12' followed by a Spitfire PR XI also of No 8 OTU. Both are PRU Blue overall with black serials and white numbers (*A. Long*).

A line-up of Mosquito T IIIs of 204 AFS at Driffield including FMO: L-VT606 and FMO: D-RR297, all part of 'A' Flight (*R.A. Hill*).

View over the tarmac at Le Bourget, Paris, on August 23 1950. Nearest is an SE 161 Languedoc F-BCUR (No 48) of Air France. The type began as the Bloch 161 designed before the war, but the first production example as the SE 161 did not fly until September 1945. In the early 1950s Languedocs were a common sight on the London-Paris run (*M.J.F. Bowyer*).

Above To the specification which led to the piston-engined Percival Provost, T.16/48, was the Handley Page (Reading) HPR 2 *ab inito* trainer. WE505, seen here at the 1950 SBAC Display, was powered by an Alvis Leonides. Only the two prototypes were built (*M.J.F. Bowyer*).

Left A Boeing B-29-90 BW photographed in Britain during a post-war training tour. 487763 of the 509th Bomb Group has black under surfaces and appears to be unarmed. The fin is red (*Ian Dewar*).

Above right A later stage in Flying Training Command coding is seen here on VS616 'O:T' of No 6 FTS. Photographed at Bramcote in 1953 (*D. Lee*).

Right A lovely shot of Harvards of No 3 FTS Feltwell in all-yellow finish with black lettering. FB:TZ is FS756, FB:TW FX249 and FB:TT KF911 (*M. Batten*).

Left Percival Prentice FC:LG-VR303 of No 22 FTS. Silver with yellow training bands (*R.O. Hepburn*).

O T-
VS616

TT
FB TW
FB TZ

8614
AIR FORCE

Right The Martin Marlin represented virtually the end of the flying-boat era, this floating example looking most satisfied with itself at Seletar in 1958 (*D.M. Stanley*).

Left General Dwight D. Eisenhower paid a brief visit to Britain just before he left NATO command for his Presidential Campaign in 1952. Lockheed C-121A-LO 48-614 ('Columbine' on its nose in yellow) taxies out here for take-off from London Airport, with the ever-popular leader aboard (*M.J.F. Bowyer*).

Left Fairey Barracuda in close-up: RJ796, a Mk III based at St Merryn. Built by Boulton Paul, the aircraft was photographed on July 21 1951 at the memorable '50 Years of Flying' exhibition at Hendon (*M.J.F. Bowyer*).

Right The Oxford passed rapidly from the scene, but some were sold to the Burmese like UB340 seen here passing through Malta (*P. Clifton*).

Right After the war Consolidated PB4Ys of the US Navy were seen in many parts of the world, and suspected of carrying out various types of 'information gathering'. This guarded machine landed at Kai Tak in 1947 (*P. Finch*).

NAVY
3

ROYAL AIR FORCE TRANSPORT COMMAND
327
24

Right Surely one of the most magic moments in the post-war aviation scene! At the 1953 SBAC Display the gigantic Saro Princess flying-boat roared across Farnborough at low level and presented a memory for all time (*M.J.F. Bowyer*).

Right The Fairey Firefly 1 remained in use for some time after the war, an FR 1A with radome is landing here (*I.H. Gannicott*).

Left Among the participants at the 1953 Coronation Review were Avenger 3s of No 881 Squadron, Royal Canadian Navy. One machine was 53322 '314' seen here taxiing out for the practice flypast mounted from Lee-on-Solent (*M.J.F. Bowyer*).

Left After the Avro York came the Handley-Page Hastings for RAF Transport Command. WJ327 wears the blue fin diamond of No 24 Squadron, and was photographed at Adelaide, Australia, in 1956 (*J.E. Badcock*).

Right Nostalgic for some, doubtless. A C-47 taking off from Dien Bien Phu. Note the red cross marking and the fin serial 292457 (*contributor not named*).

ZR 91
ZR 78

O
292

Right The Korean War might be said to have been the first of the post-war battles between the cruel and the foolish! Shown here is a Piper Cub of the South Korean Air Force, photographed in 1950 (*contributor not named*).

Left A line-up of Lebanese Savoia Marchetti SM 81s at Beirut in 1956 (*S.J. Lewis*).

Below left Mainstay of the Fleet Air Arm in the early 1950s was still the Firefly, AS 5 VT405 '292' of No 807 Squadron off HMS *Ocean* being seen here (*A. Whitly*).

Below Firefly AS 5 VX438 of 807 Squadron, HMS *Ocean*, passes close to the summit of Mount Etna, Sicily (*A. Whitly*).

Proctor IV G-AOAR presents another link with the war years although this, too, was photographed in May 1960. Now the Proctor is all but a memory mainly of the immediate post-war years. Finish—silver and blue with light blue wings, lower fuselage and tailplane (*M.J.F. Bowyer*).

In the 1950s the Royal Navy made much use of the Westland Dragonfly, licence-built Sikorsky S 51. It can now be seen to have been the first really effective helicopter. WP494 was a Mk 3 from the last production batch, and when photographed on August 23 1961 was in the hands of Ship's Flight, Lossiemouth. Finish is Sky and Dark Grey (*M.J.F. Bowyer*).

Timeless Tiger, for sure. G-APCU served as N9508 during the war with 3, 11 and 18 Elementary Flying Training Schools and after the war with 18 RFS. It was sold for civil use in October 1953—and is still very active. The finish here is silver with dark blue trim, in the photograph taken on May 7 1960 (*M.J.F. Bowyer*).

Recalling the era when the USAF Strategic Air Command rotated its Boeing B-47 Wings to Britain is this B-47E-20-LM of the 340th Bomb Wing, based in Britain in the autumn of 1959 (*M.J.F. Bowyer*).

A de Havilland Sea Venom landing on HMS *Eagle*. The nose motif and '492' identify it as from 894 Squadron (*A.W. Feist*).

A trio of Gloster Meteor T 7s of the Royal Navy, WS104 nearest. Silver finish with yellow training bands (*T. Parker*).

A Douglas C-47 in the hands of the *Force Aerienne Congolaise*. Photographed October 18 1965 (*I.E. Peddle*).

War in the Congo brought together a motley selection of aircraft such as this Douglas B-26C Invader FM-662 named 'Makasi' and photographed January 1 1966 (*I.E. Peddle*).

A North American T-28 of the *Force Aerienne Congolaise* photographed on January 1 1966 at Albertville (*I.E. Peddle*).

The end for the Lightnings with 56 Squadron. A last line-up photographed June 28 1976 at Wattisham and including one T Mk 5, two F 3s and ten F6s (*R. Athey*).